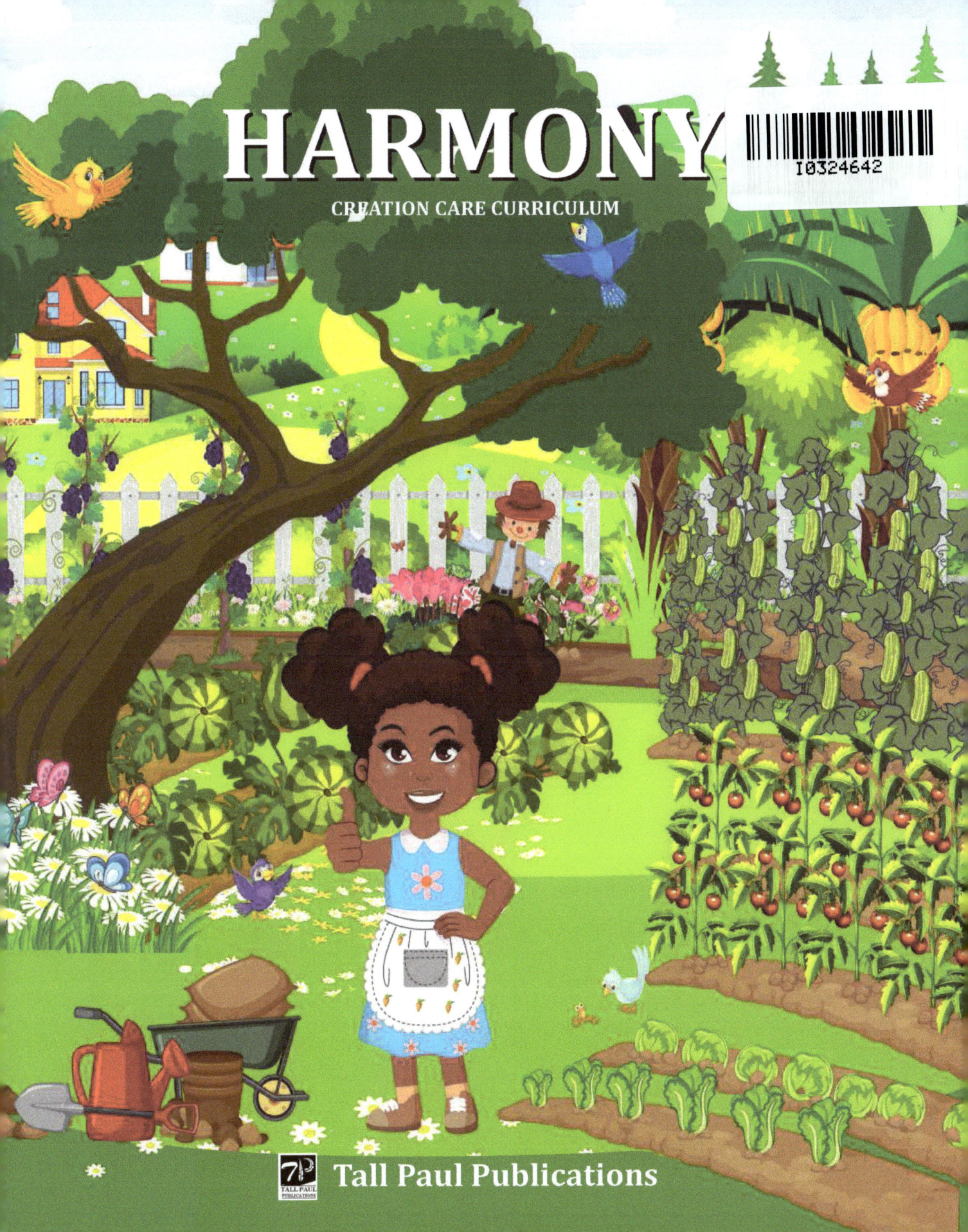

Copyright © 2021 by Crystal Paul-Watson All rights reserved.

No part of this publication may be reproduced, stored in a retrieval system, or transmitted in any form or by any means, electronic, mechanical, photocopying, recording, scanning, or otherwise, without the prior written permission of the author.

This publication is designed to provide accurate and authoritative information in regard to creation care. It is sold with the understanding that neither the author nor the publisher is engaged in rendering legal, investment, accounting or other professional services. While the publisher and author have used their best efforts in preparing this book, they make no representations or warranties with respect to the accuracy or completeness of the contents of this book and specifically disclaim any implied warranties of merchantability or fitness for a particular purpose. No warranty may be created or extended by sales representatives or written sales materials. The advice and strategies contained herein may not be suitable for your situation. You should consult with a professional when appropriate. Neither the publisher nor the author shall be liable for any loss of profit or any other commercial damages, including but not limited to special, incidental, consequential, personal, or other damages.

Harmony
Creation Care Curriculum
By Rev. Crystal Paul-Watson & Lavanda Paul

ISBN: 978-0-9835794-6-5 Paperback

Illustrations Designed By Rev. Crystal Paul-Watson & Created By Khadijah Maryam

Cover designed by Rev. Crystal Paul-Watson and Created by Rose Miller

Edited by Liz Dexter

Printed in the United States of America
Tall Paul Publications, Hartford, Connecticut

HARMONY

CREATION CARE CURRICULUM

CONTENTS

Chapter 1
THE ORIGINAL SEED   01
THE WATERMELON SHOULD HAVE SEEDS   02

Chapter 2
WHAT IS IN THE SOIL?   09
HARVESTING PEANUTS AND PLANTING THE PEA PATCH   10

Chapter 3
HAPPY WATERS   19
CATCHING RAIN WATER ON THE FARM   20

Chapter 4
THE SUN THAT RULES THE DAY   27
THE SUN BECAME A FRIEND WE DEPENDED ON   28

Chapter 5
ENVIRONMENT   33
CLEARING THE LAND WITH FIRE   34

Chapter 6
COMMUNITY: MANY AS ONE   39
AN UNUSUAL SCHOOL DAY   40

Chapter 7
MISSION – CARRYING ON THE SEED   49
A HARVEST SHARED   50

Chapter 1
THE ORIGINAL SEED

Then God said, Let the waters under the heaven be gathered together into one place, and let the dry land appear; and it was so. And God called the dry land Earth, and the gathering together of waters God called seas, And God saw that it was good. Then God said, Let the earth bring forth grass and herb that yields seed, and the fruit tree that yields fruit according to its kind, whose seed is in itself, on the earth; and it was so. And the earth brought forth grass, the herb that yields seed according to its kind, the tree that yields fruit, whose seed is in itself according to its kind and God saw that it was good.
Genesis 1:9-11

Read Scripture for Review: Mark 4:26-29

Consider which was created first, the seed or the actual food? In the creation story in Genesis, God created the earth and told it to bring forth the fruit and the vegetable plants, but God made sure to place the seed in the plants so that they can continue to reproduce. We are given the important opportunity to harvest organic seeds and plant more vegetables and fruits. We have an obligation to continue to preserve original seeds.

THE WATERMELON SHOULD HAVE SEEDS

The watermelon originated in West Africa. Watermelon seeds were preserved and found in Pharaohs' tombs in Ancient Egypt. The seeds were important then, and they are important now. God created herbs and seed-bearing plants for food as a safeguard that we wouldn't run out of food. Today many watermelons do not have the seed. Someone figured that people would rather eat seedless watermelons and then decided to genetically change the original plan. We must remember to grow organic and heirloom seeds so that we keep the original plan of God intact.

I want my grandchildren to enjoy the watermelon. I have great childhood memories of this plant. The watermelon is a beautiful fruit, and fun to eat. As a young child, I lived on a farm with my mother and father and six siblings. We raised all kind of crops, including watermelons.

The watermelon patch was by far the best. We grew all kinds of watermelons: large oval-shaped with stripes and red; then we grew the round shape with yellow inside. Both were sweet and juicy. My dad sold some and we would feast on some. It was a great family time, picking the watermelon from the patch, cutting it, slicing and eating together.

As a child, I worked in the fields, and the cotton fields were not a pleasant experience. It was hard labor. Have you heard the song "Pick a bale of Cotton"? That was a reality for many families in the 50s and 60s living in the South. Cotton was the money-crop. The watermelons grew in the cotton patches. This happened often. No one planted it; the seed just fell into the soil and grew. I remember discovering watermelon among the cotton. This was like finding a treasure. There were times I ate the sweet, refreshing melon, enjoying the break from work and the excitement in its discovery. There were other times, I would put the melon in my sack at the weighing of the cotton, and everyone was then so surprised at how much cotton they thought I had picked.

Master Gardener Lavanda Paul

FRUITS & VEGETABLES

Watermelon is one of the many plants God provided in the Garden of Eden, and it currently now has over 1,000 varieties. Can you imagine how many other plants Adam and Eve could eat in the original garden? They could eat all the fruits and vegetables that God had made except for one, obviously. There is a difference between fruits and vegetables, and I know that you can recognize the difference, or can you? For example, you may easily tell me that apples, oranges, grapes, and pears are fruits. They are sweet and juicy, and you probably love to eat them.

Vegetables, on the other hand, are often green, such as lettuce, collard greens, spinach, and broccoli. Others are red, orange, or even dark purple. Vegetables are not as sweet and juicy as fruit, but they are still important for you to eat to maintain a healthy diet. While many prefer to eat fruits, vegetables provide essential nutrients and vitamins, so your parents and guardians will insist that you eat your vegetables as well as fruits.

Do you have the answer to the question, how do we know the difference between fruits and vegetables? You will be able to tell whether a piece of produce is a fruit or vegetable by looking at the original plant. Anything that develops from a flower is considered a fruit. You pick apples, oranges, berries, bananas all from trees and bushes, but vegetables are other parts of the plant like leaves, stems, tubers, bulbs, and roots. Think about all the different types of plants and vegetables. There are about 391,000 vascular plants and 20,000 are edible, but we only eat about 20 different vegetables.

RECOGNIZE VEGETABLES

Look at the chart below. See if you can identify all of the vegetables.

Types of Vegetables	Examples of Types	Identify Vegetables
Cruciferous - Family of brassica. One of the most nutritionally dense foods we eat.	Cabbage Cauliflower Brussels Broccoli	
Marrow	Pumpkin Cucumber Zucchini	
Root	Potato Sweet potato Carrot Peanut	
Edible plant stem	Celery Asparagus	
Bulb/Allium	Onion Garlic Shallots	
Seed	Corn	
Identify the vegetable:		

Challenge: Which group would tomato, pepper and eggplant fall under? Fill out the answer in the empty box above.

Harmony Creation Care Curriculum

SEED OF THE WORD

Think about how God created the Garden of Eden. What did God do? God spoke, and it was created. God's words are like seeds. When God speaks, good things are created. Our entire beings are nourished when we hear and believe God's Word.

And He said, the Kingdom of God is as if a man should scatter seed on the ground, and should sleep by night and rise by day, and the seed should sprout and grow, he himself does not know how. For the earth yields crops by itself; first the blade, then the head, after that the full grain in the head. But when the grain ripens immediately he puts in the sickle, because the harvest has come. **Mark 4:26-29**

Reflection

Do you have a favorite scripture or Psalm? I encourage you to find your favorite scripture and write it in the lines below.

_____

_____

_____

_____

Questions for Review and Discussion:

1. Did God plant seeds to create the first garden? If not, what did God do?

_____

_____

_____

_____

2. Describe the importance of God's Word found in scripture.

3. What are some things you can do to plant God's Word in you?

4. How can we bring good things forth with our words?

Group Discussion:

Take a look at your seeds. Sort them and keep them labeled in a container or a seed packet. Think about which vegetables you would like to plant. You can also choose fruit seeds that most people consider vegetables because of the taste and nutritional value. We've identified tomatoes and eggplant as fruit, but peas, beans, and peppers would fall into the fruit category as well. Although many think of them as vegetables, beans and peas are in the family of legumes, and the pea or bean is the seed.

Choose one to three seeds that you will plant. Out of all the abundance of the seeds offered, why did you choose the one(s) you did? What type of vegetable is it? You can refer to the chart in this section. Tell those in your group about the seeds you have selected and why.

Activity 1 – Sprout Seeds

Materials needed: Paper towel, seeds, spray bottle, empty egg carton

Instructions:

1. Choose 1-3 kinds of vegetables you would like to harvest this spring and set them aside.

2. Get three wet sheets of paper towel. Put the seeds on each wet paper towel.

3. Be sure to keep them separate and label them.

4. Large seeds can also be soaked.

5. Wait and watch the seeds every day, looking for your seed to sprout. Seeds usually sprout anytime between 1-3 weeks. Keep the seeds moist by spraying them with a water bottle.

6. When you see your seed sprout, you can plant your seeds in seed starters. You can purchase seed starters in the store, or you can create your own by using a small pot and organically fused soil that is rich with nutrients to help your plants to grow.

7. Remember, check your seed package to know how deep to place the seed. Some seeds are planted one inch from the top of the soil, while others may be planted ¼ of an inch. Your seed package will guide you. Be sure not to plant your sprouts too deep in the soil.

Chapter 2
WHAT IS IN THE SOIL?

And out of the ground the Lord God made every tree grow that is pleasant to the sight and good for food. Gen. 2:9

"See I have given you every herb that yields seed which is on the face of the earth, and every tree whose fruit yields seeds to you it shall be for food." Gen. 1:29

Read Scripture for Review: Matthew 13:3-9

God is so remarkably good, and we are specially made in God's image, that in addition, God gave to us every tree, fruits, herbs, and seeds. The seeds that produce our food are our inheritance from God. As good stewards, we have a responsibility to take care of our planet and continue to use the earth to plant and grow healthy foods.

As a gardener and good steward of the earth, there are so many decisions to make, and guess what? You are gifted and empowered to make them. You have already decided on which seeds to plant. Some seeds grow better in very warm conditions, while others can flourish in more seasonal areas. Deciding how many seeds to germinate, how to provide the best soil, and where to store your seed planters or plant your garden are all decisions of the gardener.

We understand that God created the very first garden in the Middle East region. We receive hints from the rivers that were mentioned in Genesis. The Middle East has been called the Fertile Crescent and Cradle of Civilization because of the rich soil. What makes the soil rich? Water, minerals, and plant food are important variables that make the soil rich. There are microorganisms that live in the soil such as bacteria, actinomycetes, fungi, soil algae, and soil protozoa. These microorganisms interact with the other components in the soil and create rich, fertile soil ready to become plant food for crops.

HARVESTING PEANUTS AND PLANTING THE PEA PATCH

I was born into farming, and growing things is in my blood. We grew everything we needed for food. We had plenty of veggies. We grew greens, cabbage, corn, tomatoes, potatoes, and sweet potatoes, peas, squash, cucumbers, beans, onions, watermelons, and okra. My favorite crop was peanuts.

As a child, I remember harvesting peanuts. It was by far the best crop. The seeds were planted in good rich soil on our family farm, and after maturation my brothers would pull the plants up from the dirt and at the roots were the peanuts. We gathered the plants from the field and collected the peanuts. Such a fun plant for a kid. I remember my father saying we could plant only a certain amount of peanuts. It was government-regulated. I thought that was strange.

Another vivid memory from the farm was planting peas. I was about eight years old when my dad gave me the chore to plant seeds in the pea patch. It was not a pleasant job and there were three huge patches where the dirt was made ready for pea seeds. Dad instructed me to place two peas in a small hole with equal spacing. It was a boring, slow, and tedious job. He then left me to patiently follow his instructions and complete the job, but I had a better way to do it. I decided to sow more than two seeds and then quickly cover them up. I took a handful of seeds and sowed them in the soil. That was so much easier, and I finish in less than half the time. I only filled up one pea patch, though. My dad did question why all the seeds were gone, when they should have been enough for two other fields.

That year we grew so many peas, more than enough. We shared with the whole neighborhood, and I got the title pea-grower. My dad then wanted me to plant peas every year.

Master Gardener Lavanda Paul

SOIL THAT'S READY TO PRODUCE

How do you know if your soil is rich enough to produce vegetables? Consider the soil in your backyard. Do you think that it's rich with enough nutrients for you to begin a garden? Depending on where your location is, it may or may not be, but that's ok because there are things you can do to make sure your soil is ready for planting. Composting is one important thing to do. Also, many gardeners use raised garden planters that can be filled with rich soil and compost.

Compost is plant nutrients consisting of decayed/decomposed natural organic materials. You can make your own compost by collecting certain scraps from your kitchen and from your backyard. Today many gardeners make their own compost. It's made with materials such as leaves, shredded twigs, and leftover plant-based kitchen scraps. You can mix in fresh or rotten plant-based leftovers. When combined with soil, it breaks down naturally into a nutrient-rich fertilizer that feeds your crop and helps your garden grow.

CAN BE COMPOSTED	DO NOT COMPOST
Any vegetables or fruits	Meat
Banana peels	Fish
Orange peels	Butter
Coffee	Yogurt
Apple cores	Cheese
Eggshells	Milk
Onion peels	Animal fat
Twigs	Pet poop
Lettuce	These will make the compost unusable and cause disease in your soil that will destroy your plants.
Leaves	Salty or oily foods should not be mixed into the compost.

Composting is one way to respond to God's call to take care of the earth. The food we throw away in the trash is typically sent to landfills. As landfills collect our garbage, they emit or give off a gas called methane. Methane gas is produced when the trash is burned in landfills. Some of it gets trapped in the ground, which is good for the earth. However, the methane that is released into the air can be harmful to those who breathe it, and it contributes to greenhouse gases that trap heat, keeping it from escaping our atmosphere. This is one of the main causes of global warming. Can you imagine living near a landfill? The reality is, some people don't have a choice. The less waste is sent to landfills, the better we all are.

Composting is a great way to answer the call of God and take care of the earth and others. This activity adds needed nutrients to the soil and restricts the methane gas that is harmful when released in landfills, and enables it to be used in the soil to make it rich and useful for growing crops.

Activity 2 – Clean the Refrigerator

Materials needed: Container with lid

Instructions:

1. Look through your refrigerator
2. Take out any old food that needs to be cleaned out
3. Find a waterproof container with a lid
4. Separate organic leftovers that can be used as compost (use the list)
5. You can mix in eggshells, leaves, and twigs
6. Cover or close the container and put in a storage space
7. You can add leftover vegetables and organic food for weeks
8. Save the compost for your soil.

Seed of the Word – Matthew 13:3-9

Did you know that Jesus spoke a lot about good soil and poor soil? He tells us a story. By the way, Jesus' stories are called parables. This story is called the "Parable of the Sower."

"A farmer went out to sow his seed. As he was scattering the seed, some fell along the path, and the birds came and ate it up. Some fell on rocky places, where it did not have much soil. It sprang up quickly, because the soil was shallow. But when the sun came up, the plants were scorched, and they withered because they had no root.[7] Other seed fell among thorns, which grew up and choked the plants.[8] Still other seed fell on good soil, where it produced a crop—a hundred, sixty or thirty times what was sown. Whoever has ears, let them hear."

Matthew 13:3-9

The disciples asked Jesus why he spoke to everyone in parables, and Jesus explained to them that the mysteries of the Kingdom of God is given to those who really want to know and understand from a sincere heart. Jesus then explained the parable to the disciples. Read the rest of Chapter 13 in the Gospel of Matthew. The seed that is planted is the Word of God. Jesus tells of four different scenarios about what happens when a person hears the Word of God.

Sown on a path
Those who hear the message about the kingdom, but they don't understand it. Then, the evil one comes and snatches the Word that was planted in their heart.

Sown on Rocky Ground
Those who hear the message about the Kingdom and are very happy and excited, but when trouble happens because of the Word they quickly forget or fall away because they do not develop roots.

Sown Amongst Thorn
These persons also hear the word, but they become worried about life or trying to become rich, and the Word is choked and cannot produce any fruit.

Good Soil
The last type of people Jesus mentions are those who hear the message from God and understand it. The Word of God takes root in them because they take time to understand what God's Word means. The Word matures in the hearts and minds of faithful believers, and these individuals produce fruit in their character and lives that derive from the good Word of God.

What did each of the people Jesus was talking about have in common?
Answer:
They all heard the Word of God

There are many people who hear God's Word. However, it's important that you be one of the ones who value and seek to understand God's Word. It is one of the reasons God the Son Jesus Christ came to earth, lived, and taught the wisdom of God. Understanding the love and goodness of God through the Word is most important. The knowledge of God is powerful and able to transform our lives in every way. Genesis tells us that Elohim created the entire world with the Word. Everything that is good and beautiful was created with the Word of our Creator. How important it is for us to hear God's instructions and thoughts toward us because we are made in God's image. Jesus was sent to the world so that everyone could understand God's love. We can learn and understand the Word by listening to and discussing the teachings of Jesus.

The person who takes time to understand has truly recognized God as the source of wisdom, life, and goodness. When trouble comes, this person stands on the promises and understanding of God's faithfulness. Instead of choosing to worry, they choose to trust that God does take care of us and respond to our needs.

Secondly, a good heart that is ready and open to God's Word does not allow its focus to be diverted to riches and things in the world that distract. God's Word is too important to them. Instead, they spend time meditating, praying, listening, and talking to other believers. Believers also spend time doing acts of love such as caring for widows and children, feeding the hungry, and visiting and healing the sick. Believers spend

Seeds thrown on path

Seeds thrown amongst thorn

Seeds thrown amongst rocks

Seeds thrown on good soil

Seeds sprouting

time doing for others in need in the community because they follow Christ Jesus' commandment to love, also knowing that God, who has created all good things, has promised to bless those who believe. This person matures in God's Word and love and brings forth a lifetime of goodness that the Word of God produces.

Questions for Review and Discussion:

The good soil is described as soft, rich in nutrients, moist healthy soil that is ready and waiting for the seeds to be planted in it.

How can you prepare your heart to be the good soil so that the seed of the Word of God can be planted and grow to bring forth good things?

_____

_____

_____

_____

Do you recognize any people around you who have planted the Word of God in their heart? Describe some of the good things that come from one of these persons.

_____

_____

_____

_____

Why is it important for believers to know the difference between good soil and rocky, trodden, or thorny places?

_____

_____

_____

_____

Chapter 3
HAPPY WATERS

In the beginning, God created the heavens and the earth. Now the earth was formless and void, and darkness was over the surface of the deep. And the Spirit of God was hovering over the surface of the waters.

Read Scripture for Review: Genesis 1:1-2
Matthew 3:13-17, Mark 1:9-13, or Luke 3:21-22, Genesis 6-9, Exodus 14:19-25,
Numbers 20:1-11, Joshua 3:9-17

God has done miraculous work with water. Its wet, fluid, pure substance has been present even since before creation. In the beginning, when the earth did not have any form, it was the water that was present in the darkness. The scriptures tell us that God's Spirit hovered over the waters. God's Spirit has been hovering over many waters since the beginning of time, and the waters have served God in return. A fetus in the womb in a mother's belly is surrounded by water. It is when the water sac breaks, that a mother is signaled that the baby is ready to be born.

We know how important water is to the earth and to our life cycle as human beings. About 55%-65% of our human body consists of water. Within us, the liquid substance functions in many different ways, including transporting oxygen to every part of our bodies and converting food to components needed for survival. Water also regulates the body temperature and flushes the body's waste.

Just as humans need water to survive, plants depend on water as well. Plants use more water than humans. As much as 95% of plants consist of water. Remember when watering your plants that water enters through the roots and then is translocated to every part of the plant. It is the water that fills the vascular part of the plant cells which gives the plant its structure and shape. Water is filled with nutrients that plants use for food. It is so important that your plant receives the right amount of water all throughout its lifespan for it to survive, be healthy, and produce.

CATCHING RAIN WATER ON THE FARM

As a youth, I learned the importance of not wasting anything. Nothing was lost or wasted on our family farm. Instead, everything had more than one use. The farm had a way of making you think and value resources and use them wisely.

As a child, I remember catching rain water to water the garden. The rain water filled large drums, which are like huge 50-150-gallon metal containers. The rain water also filled our well. There were no faucets; instead every house had their own well, and we used containers to carry the water. There were natural springs nearby also. We didn't have what you would call running water in our home. We had to draw water from the well and carry it into the house and use it for all our needs in the home: drinking, cooking, washing, cleaning, baths and toileting. It was work, and the water was very heavy to carry.

A good rainfall was so refreshing to the crops and also fun to play in. The smell of rain was so delightful. As a child, I learned to appreciate everything the earth gave us.

Master Gardener Lavanda Paul

Activity germinates seeds

At the very beginning of the growing process, water is the key to transform the seed. The first stage of growth in the life of the plant is called the germination process. The seed must transform; germination is the process of the seed breaking open, changing to a sprout, and transforming into a seedling. Seeds can be soaked to begin the germination process. Larger seeds are easier to soak. Tiny seeds often cling together when soaked. Remember the different ways to germinate seeds: small seeds can be placed on small pieces of cut-up paper towel placed in an empty egg carton, sprayed with water mist, and covered with the top of the carton to keep in moisture, and larger seeds can be soaked in water. However, many seeds do better with darkness and should be planted directly in the soil. Check your seeds daily to see if they have sprouted. Once the seeds germinate, they will be ready to grow in the soil. It usually takes between 1-2 weeks for seeds to sprout. You can also try simply planting the seeds directly in rich, fertile soil.

Seed of the Word

Read and review scriptures: Genesis 6-9, Exodus 14:19-25, Numbers 20:1-11, Joshua 3:9-17

As we hear, understand, and accept God's love for us through Biblical stories, we are also transformed like the seed. Spending time in

discussion with others who believe helps us also to grow as a watered plant does. One of the first examples we see in the Bible regarding rainwater is the rainwater God sent to cleanse the earth from violence and evil. Noah was found to be righteous in God's eyes, and God provided refuge for his family and all the species in the Ark as they were delivered from 40 days and nights of pouring rain. God promised that he would never destroy the earth again with water, and the rainbow was given to us to see when it rains as a reminder of this covenant promise.

> **The seed germination process**
> Imbibition – water fills the seed
> - The water activates enzymes that begins plant growth
> - The seeds grow roots to access the water underground
> - The seeds grow shoots that grow toward the sun
> - The shoots grow leaves

At another time, God's chosen people, who are called the Children of Israel, needed to cross the waters of the Red Sea to escape their enemies. God commanded the waters to open up, so the people could walk through, and when their enemies came following after, God released the water, which swallowed up and destroyed their enemies.

As the Children of Israel traveled through the wilderness, they cried out to the Lord because they needed water, and God made provision. After Moses struck a rock, water gushed out. The people's thirst was satisfied when they drank water from the rock.

The Jordan River was the only thing that separated the Children of Israel from the land that God had promised they would live in and inherit. God instructed their leader Joshua, and he struck the Jordan River with his cloak; the river then parted so that the Children of Israel crossed over the Jordan River on dry land.

The New Testament also highlights God's important use of water. You can read about John the Baptist, who baptized believers in the Jordan River in the wilderness in the gospels. People came from all over to be baptized by John the Baptist. In the Methodist heritage, baptism is a choice that either a person makes when they realize and understand the work of salvation provided by God through Jesus Christ, or a choice parents make when they choose to nurture their baby in the church for growth in God until the child is old enough to come into their own faith. John the Baptist performed many baptisms of repentance in the wilderness. Even Jesus, at the start of his ministry, went to be baptized in the Jordan River by John. He did this so all believers would follow his example and be baptized, too. Read about that in the Gospels of Matthew 3:13-17, Mark 1:9-13, or Luke 3:21-22.

Baptism Waters

Could you imagine the ordinary water of the Jordan River blessed by John the Baptist used to baptize Jesus? Those must have been some happy waters. The same clear, fluid substance is also blessed to baptize any soul who repents from sin and comes to realize and believe that we are all in need of a Savior. God saw our need and responded when God sent His only begotten Son, Jesus Christ, to be our Savior. A person who chooses to be baptized or whose parents make an outward confession that they believe in Jesus Christ and accept him as their Savior accepts that Jesus died on the cross for them, and Christ forgives them of all their sins and even takes their sin away.

Jesus justifies the believer who repents from their sins, and through baptism, we are incorporated and welcomed into the family of God called the church. We are reconciled to God through Jesus Christ, who gave his life for our sins and the sins of the whole world. There is love in the happy water. Sustenance and provision made by God are within the water. There is the power to wash away all that is dirty and filthy in the waters of baptism that are blessed. The waters represent an outward agent used to symbolize the invisible inward work of the Holy Spirit that is poured into the heart and soul of the believer who is baptized. This is a beautiful gift that God has given to us, so we celebrate in the waters of baptism.

Heal Our Waters

We have a responsibility to preserve fresh water. Today the commitment extends to healing our natural water sources. Humanity has failed to keep our waters clean from pollution like plastic and toxic waste. We all must realize our responsibility to the earth and encourage one another as much as possible to help heal the waters and keep them pollution-free. One way to do that is to use less plastic. Although plastic is quickly made, it cannot decompose quickly. The residue of plastic bags, plastic bottles, Styrofoam, and plastic straws have often ended up in our water. You should also pay attention to what chemicals and oils get poured down drains, and fertilizer and pesticide sprays that you use in your garden, as they wash away with the rain and too much causes problems. This results in the endangerment of the ecosystem in the water. This is why more people are choosing to eat organic food.

One thing we all can do is choose as much as possible to not use plastic bags, straws, and containers when we order from restaurants or go to the store. If you really want to make a difference, also don't buy plastic bottled water. If fewer and fewer people are using single-use plastic, then there will be less plastic that ends up in the water. The second thing we can do is be sure to recycle all of our plastic bottles. The next time you drink a glass of water, think about how happy the clear, pure water is to still serve God's original purposes, as it has since the beginning of time.

Activity 3.1 – Water Plants

Materials needed: Spray bottle, Water bowls, water

Instructions:

If you are using raised planters, check your water in your planter pots. Be sure that your plants have sufficient amounts of water. Remember, plants receive water from the roots. It is wise to keep a plate that holds water below your pot and let it hold excess water. This way, you are sure to keep your soil moist and not dry.

Activity 3.2 – Recycling Leader

Materials needed: Garbage can, materials for making a sign

Instructions:

You can be the leader of recycling in your home. Check to be sure there is a separate place to store used recyclable bottles. If there is not one, you can create one. Find a garbage can that is not being used. Then, create a sign that reads:

Do you know what all can be recycled? Check the list below. Remember, you can remind your household members to not be wasteful. Instead, recycle. It's the responsibility of humanity to take good care of our planet.

Should Always Recycle	Don't Recycle
Paper	Loose Plastic Bags
All Plastic Bottles & Containers	Plastic shopping bags
Aluminum Cans	Plastic stretch wrap
Steel Cans	Foam Cups or Containers
Glass Jars, bottles & Containers	Egg cartons
Newspaper	Take out containers
Magazines	Drinking cups
Cardboard	Soiled Food Items
	Food soiled containers
	Soiled paper products
	Broken or sharp glass
	Fast food packaging
	Plastic Utensils

Questions for Review and Discussion:

1- Ask your parents or grandparents if they have planted unique seeds or preserved family seeds.

_____

_____

_____

_____

2- Ask your parents or grandparents about unique family recipes that may include food that they ate in their youth.

_____

_____

_____

_____

They may learn the skill of growing organic foods that can also be shared at the table in their own homes.

Seed of the Word

You have read about how Joseph, Samuel, and David were all called by God in their youth. David was not perfect, but his heart pleased God, and he was called to be the second king over Israel. God called him a "man after God's own heart" (1 Sam. 13:14). The prophet Samuel, who anointed David as King, was dedicated to God as a baby by his mother, who had prayed earnestly for a child. He lived in the temple and was twelve years old when he heard God calling his name one night. (1 Sam. 3:1) Joseph was around 17 years old when God gave him a dream that declared his leadership. (Gen. 37:5-11) I know these are only examples of boys and men, and that is because of the culture at the time of these events, but Acts 2:17 tells us that in the last days, God will pour out the Holy Spirit on all people, and both sons and daughters will prophesy. Throughout history, there are many examples of women declaring the Word of God.

God raises believers and equips many to nurture the people of God with truth and wisdom. Just as seeds are meant to be passed on from one generation to the next, the Word of God is also passed on to younger generations by those empowered by Christ's Holy Spirit. Christ Jesus' love is everlasting, and it is meant to be carried and shared with others.

God's mission in the world is to love. The love of God conquers all negativity through the power that only Godly love has. You are never too young or too old to carry on the mission and message of God's love. In contrast, God chooses many individuals in their youth to gain knowledge and wisdom and share it with others. Just as the wind carries seed to good soil, God pours out the Holy Spirit upon you so that you, too, can be empowered and guided in sharing and passing on seeds of love.

when sharing meals across cultural boundaries. Traditional Indians in India eat with their hands. They use their right hand; using their left hand to eat is a sign of disrespect. They also push the food into their mouths using their right thumb. They avoid putting their fingers in their mouth.

Chinese people use chopsticks. Learning how to hold the chopsticks will take practice. In Japan, it is customary to slurp your ramen or noodle soup when eating. It's an enthusiastic gesture that shows those you are eating with that you enjoy the food. Also, when you slurp loudly, some say it enhances the flavor of the food. In Iran and Russia, as well as many other hot countries, lunch is the main and the larger meal, and many shops close down for at least two hours. In some parts of Europe, dinner is referred to as the afternoon meal because it is also the largest meal. In South Korea, respect is shown to elders during the meal. Those who are younger make sure elders have received their food first before they begin to eat.

Each country and nation of people also eat a wide, diverse variety of food. We have learned many recipes that have brought joy to our tables from people who look different and speak a foreign language than what we are accustomed to. I know God smiles when we can appreciate the diversity of food created in other parts of the world. Cultivating and preparing meals from crops remind all of humanity that God provides through the earth all that we need to be nurtured, strong, and healthy.

Activity 7 – Harvest your Crops

Materials needed: Vegetables that have come to maturity, recipe book or family member

Instructions:

Think about the vegetables you are harvesting and what meal you will prepare with them when they are ready. Ask those in your homes about recipes that can include your harvest. Check your garden and your plants. See if there are any vegetables ready to be harvested. One way to be sure would be to research the amount of time it takes for your vegetable plants to mature. If they are ready, cut and wash your vegetables. Be sure to suggest they are prepared and become part of a dinner or lunch meal to be shared with those who live with you in your home.

Bringing the Community Together Through Mission

Now that you have participated in gardening and growing in your family garden or church youth group, think about ways you can share this great gift with others. Have you considered suggesting that your youth group begin a garden in the community or at a school? What a wonderful way to bring the community together productively and build relationships with others! Those who participate may also be encouraged to begin to create gardens in their own spaces.

On the other hand, the wind is the movement of large volumes of air in the earth's atmosphere; it's the flow of those gases on a large scale. The wind, in its activity, can shift and push. It picks up seeds and carries them both near and far. You may have seen them; some seeds fly like helicopters, while others glide on the breeze. The wind is both the vital breath of the universe and an active change agent.

Meanwhile, the breath of life is uniquely giving by God to all of humanity at the beginning of human existence. God breathed into human nostrils the breath of life, and in that instant, Adam became a self-thinking, individual soul. Free will is also a gift from God given at the beginning of creation. It is through grace, God's unmerited love, and work that all humans continue to breathe the breath of life given by God, as well as exercise the freedom to think and choose for themselves. However, all of humanity would be lost without God's guidance and love. We are created to be in a relationship and fellowship with God.

The Holy Spirit is God's character and mind moving, guiding, and working in the world. Christ Jesus manifests the life, work, and guidance of the Holy Spirit. When people choose to know God and follow Christ's plan for their lives, they live in the peace, joy, protection, and provision of the Holy Spirit. Choices that harm others and the earth contradict God's love and nature and separate those who do evil from God and their original good purpose. We learn from others how to make choices that please God and benefit the whole community.

Learning how to cultivate and grow crops is one of the essential skills that can be learned and passed on to support and care for others and take care of the earth. Still, today in many cultures, people learn how to plant and cultivate produce from the land in their youth. They participate in planting, growing, and harvesting in their families and community tribes. Others learn gardening and farming in their adult years. You are never too old or young to learn the skill of gardening.

Regardless of whether you participate in growing and harvesting crops, there is one activity that every person in the family and tribe can contribute to. That is, enjoying the harvest. Harvest is abundant when seeds are sown and cared for to maturity. It is such a blessing when all are invited to the table or eating space to commune and eat together. Coming together around a shared meal is one way to pass on knowledge and wisdom.

Different cultures have different ways of sharing a meal, and communities are nurtured through sharing. Different nationalities eat differently to nurture their community. The differences should never be frowned upon. Instead, we celebrate God's diversity when we joyfully adapt to other cultures' traditions

A HARVEST SHARED

> My community consisted of four walking-distance homes where cousins, aunts, and uncles lived. The nearby church also served as the schoolhouse for all twelve grades, with one teacher and a bus driver.
>
> My mother had three sisters, all with husbands and children, and we were very close. We rotated where Sunday dinners would be held. My dad usually cooked when dinners were scheduled at our home. Sharing was a way of life.
>
> On the farm, when someone killed a hog, it was shared throughout the community. Our gardens were more like neighborhood gardens. it was all about sharing. Everyone planted a garden, but in your abundance sharing was the way of life.
>
> *Master Gardener Lavanda Paul*

Ruach: Wind, Spirit, and Breath

In our world, where there is sometimes chaos and darkness, God's Holy Spirit brings life, wisdom, and light. We've read that in the beginning, when everything was dark, formless, and in a state of chaos, "the Spirit of God hovered over the waters." Even where there is obscurity and chaos, God's Spirit is present, offering order, life, and wisdom. The translation of "Spirit of God" in Hebrew for this verse is Ruach Elohim, or great wind. The Hebrew word Ruach has been translated as wind, Spirit, and breath. Let's explore this: breath, air, wind, and Spirit. Are they similar or all the same?

Starting with air: our air is a combination of gases. The air in our atmosphere is made up of about 78% nitrogen and 21% oxygen. Other gases in the air are argon, carbon dioxide, and fractional amounts of methane, helium, neon, krypton, and hydrogen. Our atmosphere can also hold up to about 4% of water vapor in the air. We already know that there is a unique relationship between trees and the air we breathe. Trees must take in carbon dioxide to live, and through the process of photosynthesis, they release oxygen that we need to live.

Chapter 7
MISSION – CARRYING ON THE SEED

The Lord God formed the man from the dust of the ground and breathed into his nostrils the breath of life, and man became a living soul. **Genesis 2:7**

The wind blows where it wishes, and you hear its sound, but you do not know where it comes from or where it goes. So it is with everyone who is born of the Spirit. **John 3:8**

Read Scripture for Review: Matt 8:27, 1 Samuel 16, 1 Samuel 3, Genesis 37:5-11, Acts 2:17

Focus – Extend God's love to the world through actions of care for others. Many are called in their youth to continue God's mission.

Seeds are meant to be passed on to the next generation. Gardening and planting crops are methods that nations of people have participated in nourishing the community and carrying on the heirloom organic seeds. Can you imagine a world where seeds no longer exist? That would be terrible and destructive, because seeds provide the necessary food for all people to survive. In all of our knowledge and use of technology, we must guard against patterns and behaviors that destroy and do not harness and cultivate seeds. Instead, we can work together with God and nature to preserve and pass seeds to the next generation. Communities preserve and continue the seeds by helping one another and sharing our time, energy, and gardening talents. Helping those who are older and more knowledgeable about planting also allows young people to gain insight into gardening and planting that can then be passed down to future generations.

2- What is the difference between fruitful and unfruitful work? Give examples. Tell how that relates to the city of Babel and the instructions that were given by God.

_____

_____

_____

_____

3- Describe what happened on the Day of Pentecost and why this day is so important in the history of the church.

_____

_____

_____

_____

4- Although God's people may speak in many different languages, describe the commonalities the Holy Spirit brings.

_____

_____

_____

_____

One particular year was the most important Passover year in the history of time. Jesus, God's Son, who was fully human and fully the Deity, had been the sacrificial Passover lamb. He died on the cross and rose from the dead. After he rose, we learn through scripture that he spent a good amount of days with the disciples. Eventually, Jesus ascended into the heavens, but he made a promise to the disciples; he would send them a comforter that would guide them and be with them forever.

The Day of Pentecost is the fulfillment of Jesus' promise. The disciples were together in one house when suddenly a sound like a mighty rushing wind came from heaven. The Holy Spirit filled the entire house and filled the disciples. The disciples were given the ability to speak of God's goodness and glory, each in a different language. When this happened, people in Jerusalem gathered around, and each heard their native language.

This day marks the beginning of the church, where nations of people heard and witnessed the power and true wisdom of God spoken in every language. Although the disciples spoke in different languages, the message was the same between them: God's goodness, glory, and power are real, and God has come to live within the hearts and minds of all people who are open to receive. The community of faith is endowed with the power to transform the world with the good news and work of the Kingdom.

Questions for Review and Discussion:

1- Describe what it may have been like to live in the city of Babel. Also, think about how communities can often influence individual choices.

_____

_____

_____

_____

Activity 6 – What Grows Near You?

Materials needed: Reference books or Wikipedia

Instructions:

Did you know that the United States has the most farmland over every other country in the world? Do some research on your own to discover what major crops are grown nearest to your home in the US.

Be sure to care for your garden plants and include a family member in your garden work. The five major cash crops are 1) sugarcane, 2) corn, 3) wheat, and rice is number four.

Seed of the Word: Acts 2:1-31, John 16:7

The diversity that God has created in our world is truly a celebration of beauty and uniqueness. We gain a great deal learning how other cultures have found solutions to challenges, learned how to work together, and provide crops for their communities. Although God scattered the people at the Tower of Babel and gave them different languages, this was not intended for people to hate and harm one another.

God wanted the people to be productive, working toward the original plan of replenishing and caring for the earth and one another. God always intended to come down to dwell with creation. Read Acts 2:1-31. People from every nation in the world were gathered in Jerusalem on the Day of Pentecost. Isn't that remarkable? Although the Hebrew Bible expresses how different nations of people were scattered in the first place, God engrafted a plan through the threads of time to gather the people together again, but this time to the salvation found in Jesus Christ and the wisdom of truth. The different nations of people had gathered through their seeking and obedience to celebrate the Jewish Feast of Shavuot or Weeks. The Feast of Shavuot celebrates the wheat harvest in Israel and also commemorates when God gave the Ten Commandments to the nation of Israel. It was one of the most important traditions and celebrations in Jewish culture.

Notice how God gathers the people with a feast. Since the beginning, families have gathered together to thank God for the provision of food and all necessities for life. The Feast of Passover is celebrated 50 days before the Feast of Shavuot because it is known to have been 50 days between God leading the Children of Israel out of Egypt and giving them the Ten Commandments on Mt. Sinai. The Feast of Passover commemorates the death angel passing over each door of the nation of Israel that had the sacrificial blood, the Egyptians releasing the Children of Israel from slavery, and God leading the people out of Egypt.

when it bears red fruit. Harvest for the ginseng plant begins in May and lasts through October. Hymyang is a town in South Korea and one of the few places in the world that cultivates wild ginseng. This town holds festivals once a year to celebrate the harvest. During the festival, rare ginseng root is provided to guests for a small price.

Families, tribes, and nations of people have been harvesting crops since the beginning of time. Throughout all parts of Africa, you may still see today groups of women walking with large baskets balanced on their heads and small children tied in a sack on their backs.

They balance the baskets on their heads and keep their children tied to their bodies so that their hands can remain free. These baskets are often filled with produce or other vegetables that have been harvested. Africa's major crops are potatoes, bananas, plantains, cassava, sugar, groundnuts, coconuts, soybeans, and sesame.

In India, you may see Hindu women dressed in colorful dresses and head coverings and working together to gather huge bundles of rice crops, wheat crops, corn crops, chickpeas. These are called cash crops because people can earn a profit from producing and selling a great deal. The mango is another crop that has gained India world recognition for production.

In China, you may see entire families harvesting crops such as rice, potatoes, and tea leaves. These Chinese farm workers wear large-brimmed hats to protect themselves from the heat of the sun on their face. Large baskets are also used to collect tea leaves. Much of China's land is too mountainous or arid for farming, but the eastern and southern regions are highly productive.

In Brazil, there is an indigenous tribe called Yanomami who live in villages in the Amazon rainforest. The women of the tribe gather large amounts of food from the garden. Although the men clear the garden areas at the beginning of the planting season, the women usually work in the garden, gathering nuts, fruits, and vegetables. They cultivate the garden for as long as it is fertile, while the men hunt. The Yanomami women decorate their bodies and their baskets with a red berry called an anoto. When the tribe has had a good harvest, they celebrate with a huge party by gathering large amounts of food and inviting nearby villages. The tribe has a village feast, and the women dance and sing songs at night. Sugarcane from Brazil is the largest crop in the world. Also, tomatoes, peppers, and potatoes all come from South America.

Europe is known for its cereal production. Three hundred seventeen million tons of wheat and spelt are produced in Europe, and 39% of Europe's land is cropland, where farmers grow barley, corn, alfalfa, soybean, and oats. France has the highest agricultural production of cereal, and this country also is known for producing wine, poultry, and beef.

There is a plant that is grown in South Korea that has the unusual shape of a person. They call it a holy herb. This plant is called wild ginseng and is known to strengthen the immune system, enhance cognitive and brain function, and improve mental health. It's most nutritious

and a tower that reached heaven and stretched up to God. Can humans build a skyscraper that reaches heaven's throne? We know the answer because we have built rockets and airplanes that have navigated the skies and space.

The people in the city of Babel thought that they could accomplish this goal, and they all put their minds and energy into building the tower. When God saw the city and the tower, God decided to give the people different languages. God acknowledged that the people of Babel were all motivated in the same goal, and they eventually would be successful in building a skyscraper that reached space. If they continued to build, they would finally get to a place in the sky where they would have difficulty breathing. Their work would become more harmful than productive. It is so important to God that we are using our time and energy to do things that are productive and fruitful. It is a waste of time for communities to engage in destructive and harmful endeavors.

As a result, God decided to divide the people. Different groups of people were given different languages, and the people could no longer understand one another. As a result, they dispersed to different parts of the earth. We learn a lot from this Biblical story. One lesson is that communities are compelling and can accomplish a great deal when groups of people unite to fulfill a common goal. However, as good stewards, the shared goals should always benefit those living both within and outside the community and the environment and not harm them.

AN UNUSUAL SCHOOL DAY

My early life on the farm was the best. My six siblings, mother, father and grandmother were my world. Caring and sharing is what people did in those days.

I remember a time when my aunt and her family had gotten behind in harvesting their crops. When this happened, all the children in our family would miss school until the crops were secured. This could cause great problems for us at school. Nevertheless, my mother and all of us would show up at my aunt's farm to help her harvest her crops for the day. It only took one day of everybody sharing the workload for my aunt to catch up. After the crops were tended to, we were all able to return to school.

Master Gardener Lavanda Paul

Chapter 6
COMMUNITY: MANY AS ONE

Then God said, "Let us make humankind in our image, according to Our likeness; and let them have dominion over the fish of the sea, and over the birds of the air, and over the cattle, and over all the wild animals of the earth, and over every creeping thing that creeps upon the earth." Genesis 1:26

When no plant of the field was yet in the earth and no herb of the field had yet sprung up—for the Lord God had not caused it to rain upon the earth, and there was no one to till the ground; but a stream would rise from the earth, and water the whole face of the ground—then the Lord God formed man from the dust of the ground and breathed into his nostrils the breath of life; and the man became a living being. And the Lord God planted a garden in Eden, in the east; and there he put the man whom he had formed. Out of the ground the Lord God made to grow every tree that is pleasant to the sight and good for food, the tree of life also in the midst of the garden, and the tree of the knowledge of good and evil. Genesis 2:5-9

Read Scripture for Review: Genesis 1:24-29
Genesis 2:5-18, Genesis 11:1–9, Acts 2:1-31, John 16:7

God created communities to strengthen and build those who are a part of it. The activities of farming and gardening have united communities all over the world for generations. Unique strategies for planting and cultivating crops have been passed down from generation to generation within communities. Stories and celebrations are also shared along with all other traditions in communities. Today, we have millions of communities all throughout the world. However, the scriptures inform us of a time when all the people on earth lived in one community, spoke the same language, and worked on the same goal. The city was called the city of Babel.

In the city of Babel, the people were united to build a tower that reached heaven. The people of this community must have done everything together. They planted and grew crops together and, in unity, harvested and enjoyed what they had grown. They made a name for themselves and conspired together in a mission to build a city

2- Give examples of some ways the Hebrew people during Moses and Joshua's days lived differently than others.

_____

_____

_____

_____

3- How can you ensure your environment is safe and nurturing for you to grow as a believer in Christ?

_____

_____

_____

4- What does it mean to seek first the Kingdom of God?

_____

_____

_____

Seed of the Word: Galatians 5:22-23

In the New Testament, Jesus tells us that instead of God dwelling in the land with God's people, God desires to dwell in the landscape of each heart and soul. This is a vital mission of Jesus coming to earth. The Holy Spirit is released and given to believers only after Jesus died on the cross and is resurrected from the dead. God lives within the believer, and every place we as believers put our feet becomes the dominion of God.

Scripture tells us to "Seek ye first the Kingdom of God, and God's righteousness and all other things will be added unto you" (Matt. 6:33). Jesus teaches us that the Kingdom of God is within us. We have the power to bring goodness everywhere we go. Because the Kingdom is within us, we sit together in heavenly places. We live in the Kingdom of God. Scripture also teaches us that the Kingdom is also not far from anyone: "It is in God we move and breathe and have our being" (Acts 17:28).

There are fruits of righteousness that believers possess within their hearts and souls because of the Holy Spirit's presence. The Apostle Paul explains, "But the fruit of the Spirit is love, joy, peace, forbearance, kindness, goodness, faithfulness, gentleness, and self-control. Against such things, there is no law."

(Galatians 5:22-23)

Questions for Review and Discussion:

1- Describe what a land flowing with milk and honey would be like. Be sure to tell what good qualities it would offer God's people.

_____

_____

_____

_____

Protecting your plants from insects and animals is also very important. Some gardeners use garden nets to cover their crops. These nets shield the plants from critters that would love to feast on your growing vegetables. Tall garden fences will also guard against animals such as deer. As a gardener, you must consider all of these factors and be sure your garden plants are protected and given the right environment to thrive in.

Activity 5 – Where will you Plant?

Materials needed: Your own garden!

Instructions:

Take a walk outside around your yard or on your patio. Can you find a location that will get adequate sunlight? Be sure to look up and check for trees that may block the sun. Find the perfect location to place your raised planter or garden patch. Throughout the week, be sure to check if your chosen spot is getting enough sunlight at different times of the day. Completing the chart below will help to determine exactly how much sunlight your plants will get.

Chart - Check your outside location at each hour – check the yes box if there is sunlight during each time of day.

Time	Yes Sunlight	No Sunlight
10 am		
11 am		
12 pm		
1 pm		
2 pm		
3 pm		
4 pm		
5 pm		

CLEARING THE LAND WITH FIRE

> When you live on a farm, your relationship with the land is very special. Your land is your prized possession. It is your livelihood, and the vehicle of your wealth, so you take good care of it. We had processes that my dad followed in the upkeep of the land. There was a method to clearing the land, cultivating the field, and enriching the soil.
>
> As a child, I remember clearing a corn patch: at least, my brothers did most of the clearing of the patch while I tried to help when I could. After the harvest, we cut all the corn stocks that were left in the field from the previous year. Then we gathered the stocks and put them in several piles. The best part was lighting the piles of corn stock on fire. Fire sets off excitement. There weren't many opportunities to have blazing controlled fire piles. We gathered around the fire and it was fun to watch.
>
> *Master Gardener Lavanda Paul*

Have you decided exactly where you would grow your vegetables outside? Will you use a garden patch outside, or will you stick with the raised planter or earth box? You must think about your plants' access to the sun. Remember, most plants need eight hours of direct sunlight. Make sure the place you choose provides adequate sunlight. Another thing to consider is drainage. You don't want to choose a plot of land that does not have adequate drainage. It's important to check the location for the oversaturation of water. Too much water will destroy your plants.

Having a fence around your garden bed will also protect your plants from getting too much wind. Just make sure the fence does not obstruct the sun. Lastly, it's important to limit the amounts of weeds that grow around your plants. Weeds can grow out of control and make it very difficult for your plants to grow healthy and strong. Many gardeners put down a protective layer of material called a weed guard that protects the area around the crop from weeds. Using a weed guard during planting will limit weeds growing around your vegetables.

One last thing: make sure to move your plants outdoors at the right time. You've already learned that some plants are winter hardy and can produce through the entire winter, others do not do well in cold weather or with frost, and some will be destroyed by the cold. Remember, take a look at the chart in Chapter four, and identify which vegetables can go outside at the appropriate times. You can also consult with a Farmer's Almanac for your area to gain this specific information.

Chapter 5
ENVIRONMENT

The Lord had said to Abram, "Leave your country, your people and your father's household and go to the land I will show you. I will make you into a great nation and I will bless you; I will make your name great, and you will be a blessing." (Genesis 12:1)

On that day the Lord made a covenant with Abram and said, "To your descendants I give this land, from the Wadi of Egypt to the great river, the Euphrates." (Genesis 15:18)

Read Scripture for Review: Numbers 34:1-12, Leviticus 20:24, Matthew 6:33, Galatians 5:22-23

How important is the environment that you live in? The habitat for God's chosen people and all creatures has always been very important. Not only did God make provision for Adam and Eve in the Garden of Eden, but God also called the prophet Abraham out from amongst his people and showed him the land that God had chosen for God's people. The promised land is described as a land flowing with milk and honey. Doesn't that sound nice: milk and honey? It was a land that would both nurture God's people with life sustenance and provide sweetness and goodness that would make the people of God happy.

The land for God's people was a specially chosen land that was set apart. The land has a lot to do with the well-being of people. God is deliberate and careful in choosing the perfect environment for the people of God. God gave the law of the Ten Commandments to Moses, and those who lived in the promised land all adhered to that same law. They all lived in the unique way that God instructed them to live in the land that God promised for them. During the Biblical times of Moses and Joshua, God knew that his chosen people would need their own unique environment to truly live as God wanted them to live.

Your vegetables need the right environment to thrive and grow, as well. Farmers carefully prepare the land for the crop season.

Questions for Review and Discussion:

1- According to the scripture, why did God create the sun?

_____

_____

_____

_____

2- Describe how Joshua won the battle against the Ammorites.

_____

_____

_____

_____

3- Explain how plants use sunlight compared to how people use sunlight.

_____

_____

_____

_____

4- Write down one joke you can share with your gardening group.

_____

_____

_____

_____

to continue fighting all day and into the night. Joshua knew that he would need the sunlight to shine throughout the entire battle, so Joshua had a most unusual request of God.

In the presence of God and the people of Israel, Joshua told the sun, "Sun stand thou still," and the moon stayed until the people had avenged themselves upon their enemies (Joshua 10:12-13). Joshua needed God to allow the sun to not set but shine throughout the evening until they won the battle. Scripture records, "So the sun stood still in the midst of heaven and did not hasten to go down for about a whole day, and there was no day like that before it or after it." God honored Joshua's request and helped the Israelites win the battle.

The sun also reminds us of God's mercy. God promised that the sun would continue to shine unconditionally. We remember these words, "As long as the earth endures, seedtime and harvest, cold and heat, summer and winter, day and night will never cease" (Genesis 8:22). The goodness that comes from the sun is freely given to all people regardless of their actions. This is an example of unmerited favor. It cannot be earned or achieved, or even lost. It is freely available to all people.

The sun is like a mirror to God's prevenient grace. Just as the warmth and light of the sun reaches and serves all things living, providing the source of life for all edible vegetation as well as light to our own human eyes and health to our bodies, God's mercy reaches every living soul. Prevenient grace calls out to all human beings, lighting the conscience and directing all to what is good. Although mercy and grace are present, it becomes the free will of the individual to choose to walk in the light or choose darkness.

One way of walking in the light of God is to also choose to be merciful like God. Jesus tells us, "But I say to you, love your enemies, bless those who curse you, do good to those who hate you, and pray for those who spitefully use you and persecute you, that you may be sons of your Father in heaven; for He makes His sun rise on the evil and on the good, and sends rain on the just and on the unjust" (Matt 5:44-45). We as followers of Christ have the responsibility to shine our light of love and kindness on all people. Even if they are undeserving of it, we must be examples of goodness and kindness to those who are sometimes mean, hateful, and undeserving. This is how we show that we are the children of God.

Son of God

We often see patterns in nature that reflect the spiritual work and love of God. As Christian believers, we know this truth, "For God so loved the world that he gave his only begotten Son that whosoever believes in him shall not **perish** but have everlasting life" (John 3:16). God gives Jesus who is the light of the world, and whoever walks in Christ's light and knowledge will live every day in abundance. They become spiritually nurtured and protected by Christ's Spirit.

Type	**Hardy** These are vegetables that can tolerate below freezing degrees, including frost.	**Half-Hardy** These are vegetable plants that can live through light frost	**Tender** These vegetable plants cannot tolerate frost.	**Extremely Tender** These are vegetable plants that can only thrive in temperatures of at least 65 F. (18c)
When to Plant	Winter hardy Plant first in March Or Mid to Late August	Plant these vegetables before last frost is expected	Plant well after any danger of frost.	Plant 3-4 weeks after all frost and cold weather has passed.
Vegetable Types	Onions Radishes Broccoli Cabbage Asparagus Spinach Collards Kale	Beets Carrots Cauliflower Lettuce Potatoes	Corn Beans Tomatoes	Cucumbers Melons Squash Zucchini

Frost is snow and ice. It normally begins to snow in November and can continue into May, depending on where you live in the US.

Plants are not the only ones that need sunlight to grow and thrive. Humans need sunlight as well, and not only for light. If you go outside right now and turn your face toward the sun, the warm rays feel really good on your skin, don't they? Well, at that moment when you feel warmth, natural sunlight is actually triggering the body's production of vitamin D. Vitamin D is also known as the sunshine vitamin, and it is crucial for your overall health. So many people don't get enough sunlight. Hours in so many days are used inside, working on computers and playing video games or watching television. We hide from the goodness that can only come from natural sunlight when we spend less and less time outside.

The vitamin D we receive from the sun has so many protective qualities for our bodies. For older people, it protects against inflammation, lowers high blood pressure, and helps the muscles. It also improves brain function. Our bodies are meant to be in the sun. We are advised to get at a minimum 10 to 15 minutes of sunlight daily. Getting the right amount of sunlight will help people with depression and sleep disorders. It will also help some people to lose weight, believe it or not. What better way to get the sunshine you need than to go outside and work in your garden?

Check on your seeds. Are they still moist? Did they sprout? It may be time to plant them in the soil.

Seed of the Word

We know from reading Genesis that God made two great lights, and the greater light that governs the day is the sun. The sun and moon were created to tell the time, seasons, and festival days. But, there was one very strange day, and you can read about it in Chapter 10 of the Book of Joshua. God promised the people of Israel that they would go in to possess their own land, but they first had to battle the Amorites who lived in the land of their promise. In order for the Children of Israel to win the battle, they needed

plants, and soil, to work together to sustain life. Just like water can be both life-giving and dangerous at the same time, so can the sun. Too much sun without adequate rain can kill plants and make places too hot to live. This is one reason practices that cause global warming must be taken seriously. Climate change directly affects our food production and ecosystems.

THE SUN BECAME A FRIEND WE DEPENDED ON

We got up early mornings to work and stayed out all day working the crops. During this time as a child, I was young, but we all worked. Everyone had responsibilities. My baby sister was born when I was eight years old. One of my responsibilities while out in the fields was to care for my sister while my mother worked on the farm.

Alabama gets very hot, and the summers are long. Planting season begins early spring, and the Sun plays a leading role in the success of farming. Enough rain and sunshine always made successful crops.

However, I didn't like the sun. It made working harder and my skin darker. But, I knew it was so necessary, so we made the best of it.

Before stand-up comedy became popular, we had comedy in the fields. My brothers would tell jokes all throughout the work day. Tending the crops and telling jokes was how we made work less painful and also fun. While our crops grew strong and large, we all laughed in the sun together, and I imagined the sun was laughing, too.

Master Gardener Lavanda Paul

Activity 4 – Take a look at your plants. Are they getting enough sunlight?

Materials needed: Planting chart
Instructions:

Knowing when to place your plants outside so they can benefit from direct sunlight is important. Some plants can be planted in early spring, and others need to wait until the cold and frost have completely passed. Look at the chart and identify when it is safe to plant your vegetables outside.

Chapter 4

THE SUN THAT RULES THE DAY

And God said, "Let there be light," and there was light. God saw that the light was good, and separated the light from the darkness. God called the light "day," and the darkness he called "night." And there was evening, and there was morning—the first day. Genesis 1:3-5

And God said, "Let there be lights in the vault of the sky to separate the day from the night, and let them serve as signs to mark sacred times, and days and years, and let them be lights in the vault of the sky to give light on the earth." And it was so. God made two great lights—the greater light to govern the day and the lesser light to govern the night. God also made the stars. God set them in the vault of the sky to give light on the earth, to govern the day and the night, and to separate light from darkness. And God saw that it was good. And there was evening, and there was morning—the fourth day.

Read Scripture for Review: Genesis 1:14-19, Genesis 8:22, Joshua 10, Matt 5:44-45, John 3:16

In addition to water and rich soil, plants also need sunlight to thrive. Sunlight, soil, and water combine to produce food for plants. The heat in the light causes the nutrients in the water that enters the roots and is distributed throughout the entire plant to receive the energy it needs to grow. So, light combines the gas in the air, called carbon dioxide, with the nutrients in the water to create food for plants. You may have learned about this in science class. This process is called photosynthesis. Some plants need more sunlight than others, but most plants need eight hours of sunlight to grow strong and healthy. God is the master scientist who invented photosynthesis, making a way for plants to eat and grow with only sunlight, good soil rich in nutrients, carbon dioxide gas from the air, and water. We benefit greatly, too, from this process when our plants release the oxygen we need to breathe. Purposely, God created all of earth's glory: the sun, water,

Questions for Review and Discussion:

1- Why is water so important to both our natural and spiritual lives?

_____

_____

_____

2- How did God use water to deliver the Children of Israel? Provide three different examples.

_____

_____

_____

3- What is the importance of Baptism?

_____

_____

_____

4- Describe what happens during Baptism.

_____

_____

_____